EAST ANGLIAN S*

IN THE

'THIRTIES AND FORTIES

GW00739049

Compiled by: *J. D. Mann*
Designed and Published by: **South Anglia Productions** *26 Rainham Way, Frinton On Sea, Essex, CO13 9NS.*
Telephone: (01255) 677965
Copyright: **South Anglia Productions** *1996*
Printed in England by: **Amadeus Press Limited,** *Huddersfield*
ISBN: 1-871277-37-X

Front Cover: **SOUTH LYNN June 23rd 1938.** *At the L.N.E.R. takeover former M&GN locos soldiered on for a few years until withdrawal. In this superb scene D52 4-4-0 No 047 waits with a Peterborough train.* **(M. Yarwood. Courtesy G. W. Trust)**

Back Cover: **CAMBRIDGE (Trumpington) June 1935.** *A Cambridge–Sudbury (via Long Melford) train hauled by F3 No 8069. Built 1895 withdrawn 1939.* **(J. G. Dewing)**

Inside Front Cover: **YARMOUTH (Beach) June 17th 1938.** *Ex-M&GN 4-4-0 L.N.E.R. class D53 No 06 arrives with a goods train. The engine survived until 1944.* **(M. Yarwood. Courtesy G. W. Trust)**

This Page: **LIVERPOOL STREET April 10th 1949.** *Ex-Works J69 No E8619, specially prepared by Stratford shed for station pilot duties.* **(D. A. Dant)**

Production of this special edition of EAST ANGLIAN STEAM – A PHOTOGRAPHIC TRIBUTE would not have been possible without the help of the following photographers: M. Yarwood (Courtesy Great Western Trust), L. Hanson, D.K. Jones (Courtesy B.P. Hoper Collection), C.W. and K.C. Footer, J.G. Dewing, D.A. Dant, P.J. Lynch, J.D.C. Howes (Courtesy the late L.R. Peters Collection), J.H.L. Adams (Courtesy N.R.M. York), and special thanks to L. Waters (G.W. Trust).

LIVERPOOL STREET 1938.

Class F6 N0 7003 awaits departure with a commuter service comprised of Ex-GER six wheeled stock.

(D.K. Jones, courtesy B.P. Hoper collection)

LIVERPOOL STREET 1938.

Beneath an impressive array of posters, footplate crew discuss the merits of J68 No 7041, built at Stratford in 1912. In 1939 the loco went into Government service working at Longmoor, Bicester and Faslane Gareloch.

(D.K. Jones, courtesy B.P. Hoper collection)

NEAR ABBEY MILLS
July 30th 1949.

A transfer freight makes steady progress behind a "Jinty" No 7490. It could be Monday morning judging by the lines of washing in evidence.
(P.J. Lynch)

SAFFRON WALDEN
June 24th 1938.

An ex-Great Northern railway C12 tank simmers on shed just before the outbreak of World War Two. This view typified the East Anglian branch line scene, in 1953 G5 tanks replaced these handsome 4-4-0's, these in turn being ousted by N7's in the latter 1950's.

(M. Yarwood, courtesy G.W. Trust)

TOLLESBURY
August 15th 1947.

With the war over the Kelvedon & Tollesbury Light railway returned to its rustic existence. Two boys are photographed alongside an unidentified J69 tank. The line survived only another four years.
(C W Footer)

COLCHESTER
July 6th 1948.

A study of B2 No 61616 "FALLODEN" in early British Railways livery.
(K.C. Footer)

COLCHESTER 1948.

In 1937 two "Sandringhams" were streamlined for use on the much publicised "EAST ANGLIAN" express between London and Norwich, just prior de-streamlining (1951) B17/5 No 61659 "EAST ANGLIAN" passes on the "up" working. On the right is an ex-Clacton train headed by a Gresley K2. *(K.C. Footer)*

COLCHESTER (St Botolphs)
July 6th 1948.

Five Ivatt class 2's were allocated to G.E. depots in 1951, in this rare view and obviously on trial No 6417 (46417) with L.M.S. on the tender, runs round the 5.25pm St Botolphs-Brightlingsea train.

(K.C. Footer)

NEAR COLCHESTER
July 1946.
B1 No 1272 is beautifully turned out by Norwich depot for the restored post-war "EAST ANGLIAN" express service.
(K.C. Footer)

CLACTON ON SEA
August 6th 1933.

Summer on the Essex sunshine coast brought a wide variety of motive power to Clacton. On this occasion B2 No 2814 "CASTLE HEDINGHAM" arrives with a substantial train including a four-wheeled van.

(L. Hanson)

CLACTON ON SEA
August 6th 1933.

A superb study of Gresley K2 No 4669, these engines earned the nickname of "Ragtimers".
(L. Hanson)

CAMBRIDGE (Trumpington)
June 1935.

The 2.40pm Hunstanton-Liverpool Street restaurant car express is headed by B12/3 No 8527 resplendent in L.N.E.R. green, shortly after rebuilding.
(J.G. Dewing)

RYE HOUSE
June 23rd 1949.

A B17/6 in early B.R. Livery, No 61627 "ASKE HALL" of Cambridge on an up Liverpool Street train.
(D.A. Dant)

CAMBRIDGE August 1938. B17/6 No 2804 "ELVEDON" waits in platform five with an Ipswich train.
(J.D.C. Howes the late L.R. Peters collection)

ELY
1947.

B17/6 No 1665 "LEICESTER CITY" makes a start in this broadside view, note the football emblem clearly visible under the loco name.

(K.C. Footer)

WISBECH
June 15th 1938.

Memories of the Wisbech & Upwell tramway. J70 No 7136 arrives with a goods train.
The loco has sideskirts and cowcatcher fitted for much of the route ran close to public roads.
(M. Yarwood, courtesy G.W. Trust)

MARCH (North Junction)
June 15th 1938.

Holden G4 No 8123 on carriage shunting duties. Along with 8139 these two engines by this time the only survivors of the class were withdrawn in August and December 1938.
(M. Yarwood, courtesy G.W. Trust)

MARCH
June 15th 1938.

J17 No 8228 approaches Whitemoor Yard with a goods train.
(M. Yarwood, courtesy G.W. Trust)

MARCH
June 15th 1938.

En-route to Whitemoor yard 02 No 2954 with a goods train passes the station at precisely 3.16pm.

(M. Yarwood, courtesy G.W. Trust)

MARCH (Whitemoor Junction) Another J17 No 8182 carefully negotiates the junction with a "modellers paradise" goods train.

June 15th 1938.

(M. Yarwood, courtesy G.W. Trust)

MARCH (East Junction)   G4 No 8123 in "close up". 40 engines of this class were built between 1898/1901 for suburban and branch line work.
June 16th 1938.   (M. Yarwood, courtesy G.W. Trust)

MARCH
June 16th 1935.

When the L.N.E.R. developed Whitemoor near March as the primary yard for East Anglia, one of the massive Robinson S1 hump shunters was tried out in 1931, as a result two more were built in 1932, No 6172 being released from Wath to be used at March.

(L. Hanson)

SANDY
May 1st 1937.

To the western fringes of East Anglia and a glimpse of the East Coast main line. A sunny spring morning and photographer Les Hanson captures J1 No 3003 on a local working.
(L. Hanson)

SANDY
May 1st 1937.

K3 No 2426 whisks a goods train through, note the variety of wagons.
(L. Hanson)

SANDY
May 1st 1937. A smartly turned out K2 No 4681 on a down goods.
(*L. Hanson*)

SANDY
May 1st 1937.

Streamlined A4 No 2510 "QUICKSILVER" in silver grey livery is seen on an express. The loco is one of four examples built for the "Silver Jubilee" trains.

(L. Hanson)

SANDY
May 1st 1937.

Ivatt "ATLANTIC" C1 No 4437 has steam to spare as she runs light. These popular 4-4-2's were capable of outstanding performances on expresses when required.

(L. Hanson)

Photographer Mark Yarwood was a senior manager with the Southern Electricity company and in his younger days travelled extensively throughout the British Isles. His photography was not confined exclusively to railways. Ships, churches, towns and villages, in fact just about anything of interest was recorded and catalogued with meticulous accuracy which included the exact time the shutter was released. This wonderful collection of several thousand negatives is now owned by the Great Western Trust at the Didcot Railway centre. For nine days, during June 1938, just before World War Two, Mark and his travelling companion toured East Anglia in an Aston Martin, taking photographs along the (uncrowded) way, to enhance those views already seen, the following pages contain a selection of the most interesting railway scenes from Norfolk, concentrating on the former Midland & Great Northern lines . . . .

MELTON CONSTABLE
June 20th 1938.   At 12.35pm Ex M&GN (LNER class D52) No 012 is arriving from Cromer with a passenger train.
(M. Yarwood, courtesy G.W. Trust)

MELTON CONSTABLE
June 20th 1938.

A D3 No 4345 runs in on a Lynn-Yarmouth passenger working.
(M. Yarwood, courtesy G.W. Trust)

MELTON CONSTABLE
June 20th 1938.

A down goods passes the station with Ex M&GN (LNER class J40) No 064 in charge.
(M. Yarwood, courtesy G.W. Trust)

MELTON CONSTABLE
June 20th 1938.

A closer look at J40 No 064. The M&GN used these powerful 0-6-0's exclusively for freight work except at times of crisis, built in 1896 No 064 was withdrawn in 1944.

(M. Yarwood, courtesy G.W. Trust)

MELTON CONSTABLE
June 20th 1938.

D54 No 053 arrives on an ex-Norwich passenger train.
(M. Yarwood, courtesy G.W. Trust)

MELTON CONSTABLE   Ivatt D3 No 4352 on a Yarmouth-Lynn passenger train awaiting departure.
June 20th 1938.   *(M. Yarwood, courtesy G.W. Trust)*

**SHERINGHAM
June 20th 1938.**

A Melton Constable–Cromer passenger train enters the station at 10.50am behind D52 No 012.
Almost sixty years later the scene is instantly recognisable, now part of the preserved North Norfolk Railway.
(M. Yarwood, courtesy G.W. Trust)

SHERINGHAM
June 20th 1938.

012 stands ready to depart for Cromer.
(M. Yarwood, courtesy G.W. Trust)

SHERINGHAM
June 20th 1938.

The time has moved on to 11.14am and the D52 is followed by F3 No 8088 on a down "slow" to Norwich.
(M. Yarwood, courtesy G.W. Trust)

**WEST RUNTON
June 20th 1938.**

Another M&GN scene which is virtually unchanged, although "sprinters" now call where D3 No 4355 arrives with a Melton Constable–Cromer passenger train.

(M. Yarwood, courtesy G.W. Trust)

NORTH WALSHAM Ex M&GN (LNER class D53) No 02 possibly on a Mundesley local is part of a superbly atmospheric scene.
June 18th 1938. *(M. Yarwood, courtesy G.W. Trust)*

NORWICH (City)
June 16th 1938.

Ex M&GN tank No 097 is photographed in the station yard on shunting duty.

(M. Yarwood, courtesy G.W. Trust)

YARMOUTH (South Town)
June 17th 1938.

D16/3 No 8859 makes a start with an Ipswich train in a view which encapsulates the era to perfection.

(M. Yarwood, courtesy G.W. Trust)

HALESWORTH Summer 1936.

East Anglian narrow gauge, Southwold Railway No 3 "BLYTH" is photographed for the last time having been pinch-barred out of the shed. The line had closed in 1929.

(J.H.L. Adams, courtesy N.R.M. York)